INGENUITY·SPARKLES
匠艺出新
——室内空间设计

北京纳沃佩思艺术设计有限公司 编

江苏凤凰科学技术出版社

The book *Ingenuity Sparkles* not only involves inheritance and renovation, but also summarizes the spirit and pursuit of N-Never Space. The word "ingenuity" refers to "the power of creative imagination beyond all already-existing styles". As the constructors of human dreams, designers should also be the founders of ideal life styles. Bearing this theme in mind, we shall not only be responsible for creating quality home life, but also for leading people to a healthy and green life style. Wonderful things shall be felt with great care and the art of thinking calls for forms of expression. Design is just the work which perfectly presents the nice inner feelings through the techniques of art creation. Therefore, art requires not only "ingenuity" but "sense" or "artistic conception".

Following the spirit of "ingenuity sparkle", N-Never Space inherit traditions and classics, and create many new design concepts on the basis of artistic thoughts and professional techniques. We seek to change while inheriting and attempt to innovate while changing. Originality and ingenuity are therefore held in our innovation. "Every single step forward is difficult for artists although it is well within reach." Under an almost harsh standard, we have been in pursuit of the ideal perfection in our hearts. We will advance courageously without stopping like before.

Everybody has his definition of perfection in his heart. Nowadays, various concepts are changing rapidly. People have also refreshed their definition of quality home furnishings: they shall be comfortable, practical and can meet the requirements for being people-oriented; they shall be exquisite and luxurious and particular care should be devoted to their materials and workmanship; they shall reflect profound cultures and humanistic concerns and shall be sustainable...The demands and concepts become increasingly personalized. This external factor pushes us forward to weed out the old to bring forth the new and constantly optimize our design. Internally, globalization and the fusion of different cultures arouse our unprecedented passion for design. However, innovation doesn't mean blindly making changes. Instead, it shall be carried out on the basis of inheritance and development, and must contain ideas. Aside from that, it shall also rebuild itself according to its features so as to cater for the upcoming changes. With this yearbook which is filled with the essence of N-Never Space, we intend to simply sketch the outline of the changes of the design styles. Meanwhile, we see this book as a new start from which we will constantly innovate, keep forging ahead and provide mutual encouragement for our industry.

The book selected part of N-Never Space's works since 2013 for the purpose of recording and sharing. It contains the leisure of comfortable life, the fashion of modern life style, the historical and cultural sediment caused by the shock and fusion of multi-cultures, and traditions and classics which become even fresher with the passage of time. The four parts compose the framework of the book and are also the circulating process of design. Only after experiencing the sediment of tradition and classics and understanding the profoundness of culture and the changes of history can we steer the changing multi-cultural fashion, so as to create comfortable and quality furnishings for home life, and vice versa. Comfortable living environment is our ultimate goal. Based on this, the seemingly variable fashion constantly leads people to pursue culture and art. We believe that only the unique, characteristic and humanistic designs can finally be inherited as classics.

Thus "Fashion passes, style remains".

前言
Foreword

《匠艺出新——室内空间设计》是纳沃正式面向大众推出的一本设计年鉴。"匠"取义于设计师是人类理想家园的构建者","艺"取义于技艺、艺术，设计是一种需要专业技艺并带有目的性的艺术创作。"匠艺出新"想要表达的是一种继承和创新，是在继承传统和经典的基础上以艺术思想和技艺创造出的一种新的设计表达，也是纳沃的一种精神和追求，代表着纳沃一直以来不曾停歇，追求创新和变化的脚步。

纳沃是一群年轻、热情、有朝气、有梦想的设计师的家园，在这里，有对于设计的执着与追求，有对于未来生活的憧憬与思考；在这里，纳沃将以专业的设计来为每个人量身定制适合自己且独一无二的优质家居生活。然而，对于优质家居生活的定义其实一直是在变化的：要舒适实用，符合人本需求；要豪华，讲究材质和做工；要时尚，展现独特魅力和个性；要有文化底蕴；要体现人文关怀；要环保可持续……标签一个一个地加，大家的观念一天天地变化，这就必然促使与之相应的设计也要不断地变化。然而变化的思路并不应该是一味地为了追求变化而变化，而是应该有传承和发展地变化。我们也试图用这本浓缩精华的书来简单勾勒一下设计风格的变化脉络。

这本书是关于设计的一个小结，甄选出了一部分室内设计作品来进行记录和分享，内容涵盖了舒适怡人的休闲、现代时尚的摩登、多元素多文化融合后带有文化关怀和沉淀的典藏以及经过时间细细雕琢后沉淀下来的传统四种不同风格。这四个部分是这本书的骨架，也是设计的一个轮回，只有继承了传统与经典的沉淀、体会到文化的底蕴才能把握多元素不断变换的时尚，从而创造出舒适宜人的家居环境。反之亦然，只有舒适怡人、曾经引导过时尚潮流并带有人文关怀的设计才能够最终作为经典被传承下来。

目录 Contents

006	**01 怡然自得的情怀** Leisurely Sentiment	088	**02 时尚潮流与变化** Fashion and variation
010	怀特假日 Isle of Wight Holiday	092	奥斯卡影像 Oscar Film
022	绿野仙踪 Wizard of Oz	106	色之语 Language of Color
030	西班牙海岸 Spanish Beach	118	资·味 Bourgeois Sentiment
046	圣地亚哥 San Diego	128	现代都市 Modern Metropolis
058	优雅绅士 Elegant Gentleman	146	时尚先锋 Fashion Pioneer

156	**03 文化与美的感知** Sense of Culture and Beauty	**244**	**04 艺术的沉淀** Sediment of Art
160	蜕变 Transformation	**248**	甜美法兰西 Romantic France
174	融汇 Integration	**268**	美国骑士 American Knigh
202	典藏 Reservation	**284**	乔治亚风情 Georgia Taste
		302	金色棕榈滩 Golden Palm Beach

01 怡然自得的情怀
—— Leisurely Sentiment

"休闲才是一切事物环绕的中心。"

——亚里士多德

"Leisure is the center-point about which everything revolves."

- Aristotle

因为有了阳光，世界才有了生机；因为有了阳光，内心才有了舒畅。有氧的生活状态是现代人的追求，其实心灵也需要氧气，怡然自得就是这样的一种状态，它是精神层面的一种追求，贴切一点的话也可以称之为休闲。休闲已成为这个时代重要的特征之一，它是寻找快乐，亦是生命的意义，它可以使人们保持内心的安宁，获得更多的幸福感。亚里士多德曾说："休闲是一切事物环绕的中心"。

Sunshine brings vigor and vitality to the world, and offers comfort to people's hearts. Aerobic life is the pursuit of modern people and the soul also needs oxygen. Ease is such a state of mind and a spiritual pursuit, which also can be put as, to be more exact, leisure. Leisure has become an important feature of this era, whose essence is to seek happiness and the meaning of life. It enables people to keep inner peace and enjoy more happiness. Just as Aristotle once said, "Leisure is the center-point about which everything revolves."

怀特假日　Isle of Wight Holiday

"雾雨弥漫在海面，透出曙色一线。"

——约翰·梅斯菲尔德

"A gray mist on the sea's face, and a gray dawn breaking."

- John Masefield

绿野仙踪　Wizard of Oz

"画是无言之诗，诗是有声之画。"

—— 西蒙尼特斯

"Painting is silent poetry, and poetry is a speaking picture."

- Simonides

西班牙海岸

"让我们享受人生的滋味吧，如果我们感受得越多，我们就会生活得越长久。"

——纳阿娜托尔·法朗士

Spanish Beach

"Let us enjoy life; we shall have greatly lived if we have greatly loved."

- Anatole France

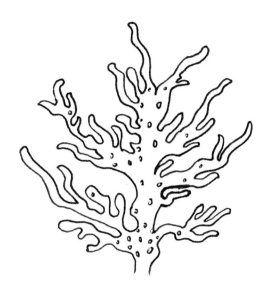

圣地亚哥　San Diego

"优雅比美丽更富有魅力。"

——拉尔夫·沃尔多·爱默生

"Grace is more beautiful than beauty."

- Ralph Waldo Emerson

优雅绅士 Elegant Gentleman

"伟大著作是由其风格和内容来衡量的,而不是其语法的修饰和多变。"

——马克·吐温

"Great books are weighed and measured by their style and matter, and not by the trimmings and shadings of their grammar."

- Mark Twain

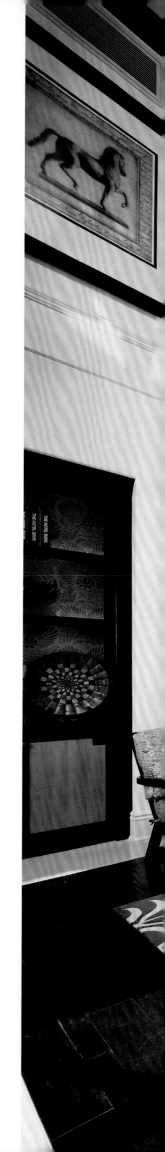

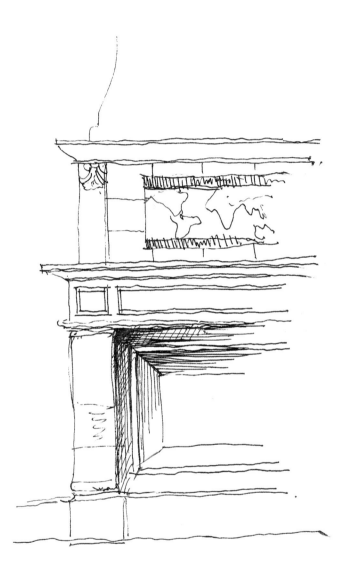

02 时尚潮流与变化
—— Fashion and variation

"时尚会过去,但风格永存。"

——可可·香奈儿

"Fashion passes, style remains."

- Coco Chanel

时尚就是在特定时间段内率先由少数人尝试，而后来为社会大众所崇尚和效仿的生活方式，是生活在这个时代的人们心里吹出的风。它不是形式，而是内容；它带给人的是一种愉悦的心情和优雅、纯粹、不凡的美妙感受；它赋予人们不同的气质和神韵，能体现不凡的生活品位；它在诠释精致中展现出十足的个性。

它总是新的、闪亮的。它源于激情，因为它总是不厌倦地做着最前沿的尝试。

Fashion refers to a life style which is firstly tested within a specific period of time by a few people and soon afterwards adored and followed by the public. It is a wind blowing from people's inner world. It's not a matter of form but of content. It brings people pleasant, graceful, pure and extraordinary feelings. It also endows people with distinguishing temperaments and romantic charms and uncommon tastes of life. Its specific features are fully expressed through the interpretation delicacy.

Fashion is always new and glittering. It's originated from passion because it has been indefatigably trying on the cutting edge of innovation.

奥斯卡影像　Oscar Film

"生活是美好的，但它缺少形式，艺术的目标正是弥补这种形式。"

——让·阿努伊

"Life is very nice, but it lacks form. It's the aim of art give it some."

- Jean Anouilh

色之语 Language of Color

"最适合你的颜色，才是世界上最美的颜色。"

——可可·香奈儿

"The best color in the whole world is the one that looks good on you."

- Coco Chanel

资・味　Bourgeois Sentiment

"哪怕沥青覆盖了整个地球，绿草迟早会冲破障碍茁壮成长。"

——伊利亚·爱伦堡

"You could cover the whole world with asphalt, but sooner or later green grass would break through."

—Ilya Ehrenburg

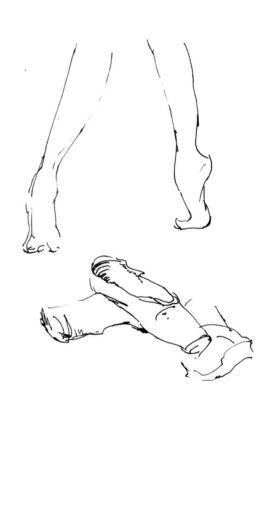

现代都市　Modern Metropolis

"设计是一种永恒的挑战，它要在舒适和奢华之间，在实用与梦想之间取得平衡。"

——唐娜·凯伦

"Design is a constant challenge to balance comfort with luxe, the practical with the desirable."

- Donna Karan

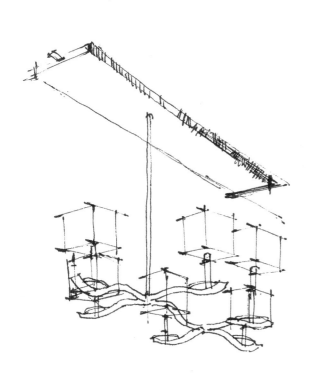

时尚先锋　Fashion Pioneer

"时尚是永无止境的变化，只要让我们能从中感到兴奋就可以。"

—— 乔治·阿玛尼

"Fashion has an insatiable appetite for change, for the new and for the innovative. Anything goes, as long as it's exciting."

- Giorgio Armani

03 文化与美的感知
—— Sense of Culture and Beauty

"文化的对撞开启了对美的感知。"

——拉尔夫·沃尔多·爱默生

"Culture opens the sense of beauty."

- Ralph Waldo Emerson

文化开启了我们对美的感知，文化的发展赋予了我们对美的审视和欣赏能力，不同民族有不同的文化，不同的文化产生出不同的美。

每一枝花都有不一样的姿态和色彩；每一个音阶都传达出不一样的情感。然而，一枝花无法构建出缤纷的海洋，一个音符无法演绎出优美的旋律。不同的文化之间相互碰撞、并存、融合，才构成了一个丰富多彩、迷人动听的世界……

Culture opens our sense of beauty and endows us with the ability to survey and appreciate beauty. Different nations have different cultures, which give birth to different beauties.

Each flower has its unique posture and color; each musical scale tells a different touching story. However, neither can a single note express a beautiful melody, nor can a single flower create a colorful sea. A colorful, rich, charming and euphonious world can be constructed only through the collusion, co-existence and fusion of different cultures.

蜕变　Transformation

"想要无可取代，就必须时刻与众不同。"

——可可·香奈儿

"In order to be irreplaceable, one must always be different."

- Coco Chanel

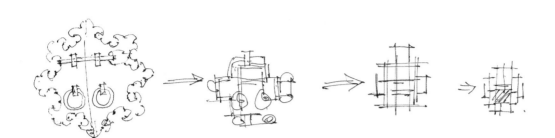

融汇　Integration

"创造，就是不断制作经典。"

"Creation is continuously producing classics."

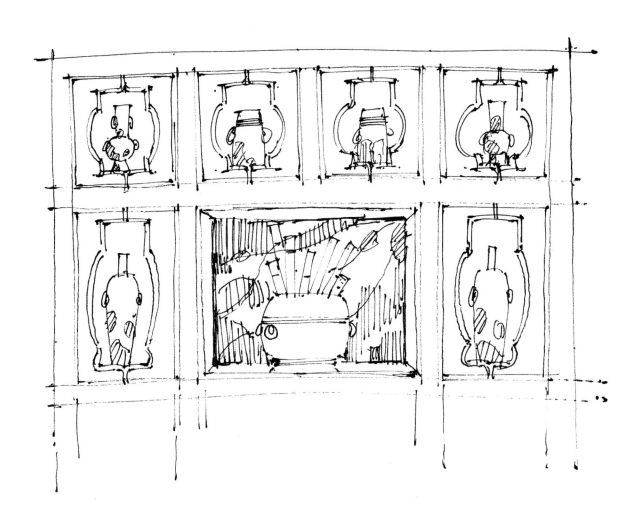

典藏　Reservation

"我喜欢并习惯了与变化的东西保持着距离，这样才会知道什么是最不会被时间抛弃的准则。"

——《西雅图不眠夜》

"I love and am used to keeping a distance with those changed things. Only in this way can I know what will not be abandoned by time."

- "Sleepless in Seattle"

04 艺术的沉淀
Sediment of Art

"艺术是永恒的，时间则是瞬息即逝的。"

——朗费罗

"Art is long, and time is fleeting."

- Longfellow

艺术的沉淀是一种文明的传承，在漫长的人类发展与社会进步过程中，不同的民族孕育出了迥异的文化，不同文化的传承与发展在历史的长河中为我们留下了众多璀璨夺目的瑰宝。

当我们轻轻走近，小心翼翼地翻开历史的扉页，沿着文化的脉络一路探寻、仔细聆听、细细品味，在感受了那呈现在眼前的极致纯粹与难忘经典之后，才不得不由衷地感叹："经典不衰，历久弥新"。

The sediment of art is a sort of civilization inheritance. During the long course of human development and social progress, different nations breed sharply different cultures. The inheritance and development of different cultures left us a lot of resplendent treasures in the long history.

We gently approach and carefully flip the open the door. Then we explore in cultural context and attentively listen and taste. Only after experiencing the extreme pureness and unforgettable classics in front of our eyes can we sigh unfeignedly that classics become even charming with the passage of time.

甜美法兰西　Romantic France

"奢侈就必须舒适，否则就不是奢侈。"

——可可·香奈儿

"Luxury must be comfortable, otherwise it is not luxury."

- Coco chanel

美国骑士　American Knigh

"回忆是层层涟漪中出现的梦。"

——H. 桑德伯格

"Remembering is a dream that comes in waves."

- H. Sandburg

乔治亚风情　Georgia Taste

"创造者才是真正的享受者。"

——富尔克

"The one who creates is the one who truly enjoys."

- Fulk

金色棕榈滩　Golden Palm Beach

"优雅不是要传达低调，而是要抵达一个人非常精华的层面。"

—— 克里斯汀·拉克鲁瓦

"Elegance is not to pass unnoticed but to get to the very soul of what one is."

- Christian Lacroix

图书在版编目（CIP）数据

匠艺出新：室内空间设计 / 北京纳沃佩思艺术设计有限公司编. -- 南京：江苏科学技术出版社，2014.6
 ISBN 978-7-5537-3186-5

Ⅰ. ①匠… Ⅱ. ①北… Ⅲ. ①室内装饰设计－作品集－世界－现代 Ⅳ. ①TU238

中国版本图书馆CIP数据核字(2014)第095267号

匠艺出新——室内空间设计

编　　　者	北京纳沃佩思艺术设计有限公司
项 目 策 划	凤凰空间 / 仝 欢
责 任 编 辑	刘屹立
特 约 编 辑	王若冰　李媛媛

出 版 发 行	凤凰出版传媒股份有限公司
	江苏凤凰科学技术出版社
出版社地址	南京市湖南路1号A楼，邮编：210009
出版社网址	http://www.pspress.cn
总 　经 　销	天津凤凰空间文化传媒有限公司
总经销网址	http://www.ifengspace.cn
经 　　　销	全国新华书店
印 　　　刷	北京博海升彩色印刷有限公司

开　　　本	965 mm×1270 mm 1/16
印　　　张	20
字　　　数	50 000
版　　　次	2014年6月第1版
印　　　次	2014年6月第1次印刷

标 准 书 号	ISBN 978-7-5537-3186-5
定　　　价	328.00 元（精）

图书如有印装质量问题，可随时向销售部调换（电话：022-87893668）。